MULTIPLE CHOICE
O LEVEL

PHYSICS

by
A.G. Harmston

CELTIC REVISION AIDS

CELTIC REVISION AIDS

Lincoln Way, Windmill Road,
Sunbury on Thames, Middlesex.

© C.E.S. Ltd.

•

First published in this edition 1979

ISBN 017 751182 6

Printed in Hong Kong

CONTENTS

INTRODUCTION

Each question or statement is followed by five possible answers or completing statements labelled A, B, C, D, and E. Write the letter which indicates the best answer.

Should you make a careless mistake or apply illogical reasoning, you will most likely arrive at one of the incorrect responses. For this reason, you must check your answers very carefully.

The first half of this book contains questions that are grouped according to major divisions in physics. In this way you may test yourself in a particular field. The second half of the book consists of a number of tests, each covering a variety of topics. Maximum benefit will be obtained if these tests are worked under examination conditions, spending not more than, say, 1 hour on one test.

Besides testing recall of facts and reasoning ability, it is hoped that these questions will form the basis of many worthwhile discussions.

Note: If necessary in any questions, you may assume that the acceleration due to gravity, g, is approximately equal to 10 ms^{-2} or 10 N kg^{-1}

1.1

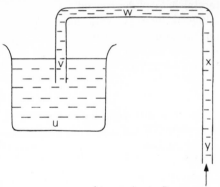

Atmospheric Pressure

A siphon is being used to empty water from the beaker above.   Pressure will be greatest at

A   point u
B   point v
C   point w
D   point x
E   point y

1.2   A metre rule, weighing 1 N, is pivoted on a knife-edge at the 30 cm mark.   If a 3 N weight is hung from the 10 cm mark, then to balance the rule, a 4 N weight must be hung at the

A   10 cm mark
B   15 cm mark
C   40 cm mark
D   45 cm mark
E   75 cm mark

1.3   A child pushes a toy off the edge of a table. Just before the toy hits the floor,

A   its kinetic energy and potential energy will both have increased
B   its kinetic energy and potential energy will both have decreased

C its kinetic energy will have increased and its potential energy decreased

D its kinetic energy will have decreased and its potential energy increased

E neither its kinetic energy nor its potential energy will have changed

1.4 A student was told to plot the total extension of a copper wire for different loads applied to the end of the wire. Which of the following graphs did he obtain?

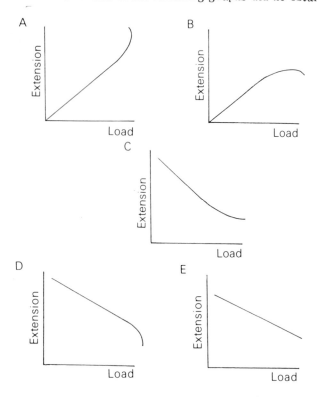

1.5 If air resistance is ignored, a stone dropped from the top of a 20 m building will hit the ground after about

    A 1 s
    B 2 s
    C 3 s
    D 4 s
    E 5 s

1.6 Two solid, rectangular blocks X and Y have the dimensions and masses shown below. Which of the following statements is true?

X: mass = 150 g

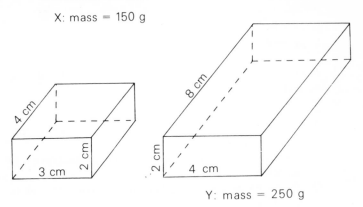

Y: mass = 250 g

    A the volume of X is greater than that of Y and the density of X is greater than that of Y
    B the volume of X is greater than that of Y and the density of X is less than that of Y
    C the volume of X is less than that of Y and the density of X is greater than that of Y
    D the volume of X is less than that of Y and the density of X is less than that of Y
    E we cannot conclude anything about the densities because we do not know the materials of which X and Y are made

1.7 Jack observes that his grandfather clock is losing time. To adjust the clock so that it will keep the correct time he should

A   increase the mass of the pendulum bob
B   decrease the mass of the pendulum bob
C   lengthen the pendulum
D   shorten the pendulum
E   do something other than the choices listed

1.8 The principle behind the operation of a hydraulic press was formulated by

A   Archimedes
B   Boyle
C   Newton
D   Pascal
E   Press

1.9 A sharp knife is more effective than a blunt one because

A   the velocity ratio of the knife is reduced
B   the cut made is narrower and less material has to be removed
C   friction is reduced because the contact area is smaller
D   the force is increased because the contact area is smaller
E   the pressure is increased because the contact area is smaller

1.10 Which graph best represents the variation of efficiency of a pulley system with load?

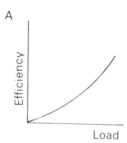

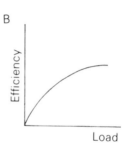

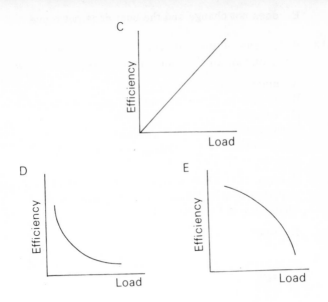

1.11 A helium–filled balloon rises in the air but eventually there is a limit to how high it can rise because

A the air is less dense higher up and the weight of air that is displaced will be less

B there is less wind higher up and so the balloon cannot continue to rise

C the density of air decreases with altitude and the helium leaks out

D the potential energy of the balloon cannot exceed its initial kinetic energy

E Newton's second law applies

1.12 A piece of soap is fixed to the rear of a toy boat which is then floated in water. As the soap dissolves, the surface tension of the water behind the boat

A increases and the boat moves forwards

B increases and the boat moves backwards

C decreases and the boat moves forwards

D decreases and the boat moves backwards

E does not change and the boat does not move

1.13 Which one of the following does not depend on how large or how small a sample of material you have?

A mass
B weight
C volume
D density
E surface area

1.14 A force of 100 N is needed to lift a load of 50 N using a pulley system whose velocity ratio is 4. The efficiency of the system is

A 0·5%
B 8%
C 12·5%
D 50%
E 80%

1.15 A small boy is standing in the corridor of a train that is travelling at 100 kilometres per hour. He jumps upward to a height of about 30 cms. When he lands his feet should touch the floor

A a long way ahead of where he was standing
B a short way ahead of where he was standing
C exactly where he was standing
D a short way behind where he was standing
E a long way behind where he was standing

1.16 A model car travels on either a straight track or a circular one. Under which of the following conditions MUST its acceleration be constant (not zero)?

A straight track: constant speed of $0·50\,\mathrm{m\,s^{-1}}$
B circular track: constant speed of $0·50\,\mathrm{m\,s^{-1}}$
C straight track: speed of $0·50\,\mathrm{m\,s^{-1}}$ increasing to $0·60\,\mathrm{m\,s^{-1}}$ in 1·0 s
D straight track: speed of $0·50\,\mathrm{m\,s^{-1}}$ decreasing to to $0·40\,\mathrm{m\,s^{-1}}$ in 1·0 s

E   circular track: speed of $0 \cdot 05\,m\,s^{-1}$ increasing to $0 \cdot 60\,m\,s^{-1}$ in $1 \cdot 0\,s$

1.17

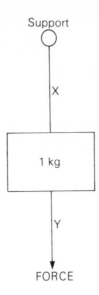

Support

X

1 kg

Y

FORCE

Two identical threads X and Y are attached to a 1 kg mass which is suspended as shown above.   The applied force may be a sudden jerk or a slow, steady pull.   Which of the following statements is true?

A   a sudden jerk breaks X: a slow, steady pull breaks Y

B   a sudden jerk breaks Y: a slow, steady pull breaks X

C   a sudden jerk or a slow, steady pull breaks X

D   a sudden jerk or a slow, steady pull breaks Y

E   X and Y are equally likely to break, no matter whether the force is a sudden jerk or a slow, steady pull

1.18 The maximum weight of an object that can be supported by two strings attached to the same point of the object, making an angle between them of 120° and each capable of sustaining a tension of 20 N, is

A 10 N
B 20 N
C $20\sqrt{2}$ N
D $40\sqrt{2}$ N
E 40 N

1.19 An astronaut is walking on the moon where the acceleration due to gravity, g, is one-sixth that on earth. Compared with his mass and weight on earth, on the moon

A his mass and weight are both less
B his mass is less, but his weight is the same
C his weight is less, but his mass is the same
D his mass and his weight are the same as on earth
E his mass and weight may differ, but not enough information is given

1.20 A sprinter starts his run by pushing back hard against the earth. The force with which the earth pushes back on the sprinter is

A greater
B zero
C the same
D a little less
E very much less

1.21 A boy weighing 300 N sits 0·50 m from one end of a see-saw which is 3·0 m long. How far from the centre should his sister, who weight 250 N, sit if the two children are to balance each other ?

A 0·30 m
B 0·66 m
C 0·84 m
D 1·20 m

E 2·0 m

1.22 When sliding a heavy wardrobe across a floor, it is
very helpful if a rug is placed beneath the wardrobe.
This is because friction is reduced since

A the total weight will be increased
B there is less friction between the floor and rug
C the rug will not be as smooth as the base of the
wardrobe and therefore its surface area will be
greater
D the centre of gravity of the wardrobe and rug will
be higher than that of the wardrobe alone
E the rug contains many pockets of air and this helps
the wardrobe to glide over the floor

1.23 Work is normally measured in units of

A joule
B watt
C kilogram
D newton
E coulomb

1.24 If you jumped a metre or more into the air from a
hard surface, it would be better, i.e. less pain
when you land to

A keep your legs straight as you are then more
likely to bounce
B keep your legs straight since your potential energy
would then be greater and so your kinetic energy
will be less
C allow your knees to bend because your potential
energy, and therefore your total energy, will be
be less
D allow your knees to bend since this would lessen
the rate of deceleration and thus the stopping force
will be reduced
E ignore the above advice because it makes no
difference whether you bend your knees or keep
your legs straight

Questions 1.25, 1.26 and 1.27 refer to the following diagram. The bob of a simple pendulum is released at point f and swings through point g to point h.

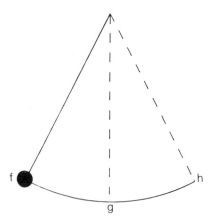

1.25 The kinetic energy of the bob will be a maximum at

    A   points f and h
    B   point g
    C   somewhere between points f and g
    D   somewhere between points g and h
    E   points f, g and h, since the kinetic energy is constant

1.26 The velocity of the bob will be a maximum at

    A   points f and h
    B   point g
    C   somewhere between points f and g
    D   somewhere between points g and h
    E   points f, g and h, since the acceleration is constant

1.27 The momentum of the bob will be a minimum at

    A   points f and h
    B   point g
    C   somewhere between points f and g
    D   somewhere between points g and h
    E   points f, g and h, since the momentum is constant

1.28 It is harder to walk on ice than on a concrete path because

A the ice is harder than the concrete
B the density of the ice is less than that of the concrete
C the ice is wet and the concrete path is dry
D the friction between the ice and the soles of our shoes is very low and our feet tend to slip backwards whereas the friction between a concrete path and our shoes is high
E the weight of the body is distributed over the whole of the ice so that the pressure exerted on the ice is less than on the concrete path

Questions 1.29 and 1.30 refer to the following graph. This shows how the velocity of a cyclist travelling along a straight road changes over the equal time intervals f-g, g-h, h-i, i-j, and j-k.

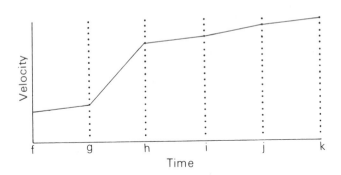

1.29 The cyclist's acceleration was greatest during the time interval

A f-g
B g-h
C h-i
D i-j
E j-k

1.30 He covered the greatest distance during the time interval

    A f-g
    B g-h
    C h-i
    D i-j
    E j-k

2.1 Which of the following equations correctly relates a temperature F in degrees Fahrenheit and a corresponding temperature C in degrees Celsius ?

A $\quad F = \dfrac{9}{5} \times C + 32$

B $\quad F = \dfrac{9}{5} \times C - 32$

C $\quad C = \dfrac{9}{5} \times F + 32$

D $\quad C = \dfrac{9}{5} \times F - 32$

E $\quad$ none of these equations is correct

2.2 Which type of thermometer would it be best to use to measure the temperature of a sample of mercury of volume $1 \cdot 0 \times 10^{-6} \, m^3$?

A electrical resistance
B mercury-in-glass
C alcohol-in-glass
D thermocouple
E pyrometer

2.3 A piece of ice of mass 1 kg is heated until it has just completely melted. This requires a total heat energy of $339 \cdot 4$ kJ. If the specific latent heat of fusion of ice is 336 $kJ\,kg^{-1}$ and the specific heat capacity of ice is $1 \cdot 7$ $kJ\,kg^{-1}K^{-1}$ then the ice must have originally been at a temperature of

A $\quad 0^{o}C$
B $\quad -1^{o}C$
C $\quad -2^{o}C$
D $\quad -3^{o}C$
E $\quad -4^{o}C$

2.4 Pure naphthalene has a freezing point of $80^{o}C$ and a boiling point of 218°C. Its melting point is

A below 80°C

B 80°C

C between 80°C and 218°C

D 218°C

E above 218°C

2.5 On a sunny day, a driver parks his car with all the windows closed. When he returns to the car, he finds it is unbearably hot inside. This is mainly because the infra-red rays from the sun

A are absorbed by the glass windows and the interior heats up by convection

B pass through the windows but cannot escape because of total internal reflection

C pass through the windows and are absorbed by the seats which then emit long rays that cannot pass through the glass

D are focussed into the interior by the windows which act as lenses

E pass through the windows and out the opposite side, heating the interior as they pass

2.6 When you stir tea or coffee, a metal spoon always feels hotter than a plastic one because

A the metal is at a higher temperature

B the metal is a better thermal conductor

C the metal is a better thermal radiator

D the metal has a much greater mass than the plastic

E moisture condenses more rapidly on the metal

2.7 The specific heat capacity of a substance is a measure of the heat energy needed to raise the temperature of

A 1 m^3 of the substance by 1°F

B 1 m^3 of the substance by 1 K

C 1 kg of the substance by 1°F

D 1 kg of the substance by 1 K

E 1 m^3 of water by 1 K

2.8 On Guy Fawkes night, you may safely hold a 'sparkler'. The sparks do not burn you because

   A the temperature is not very high
   B moisture from the hand forms a protective barrier
   C the mass of a spark is very small
   D the hand is a poor thermal conductor
   E the sparks are in contact with the hand for a negligible time

2.9 A disc of copper has a circular hole in it. When the disc is heated uniformly

   A the circumference of the disc increases and the circumference of the hole increases
   B the circumference of the disc increases and the circumference of the hole remains unchanged
   C the circumference of the disc increases and the circumference of the hole decreases
   D the circumference of the disc remains unchanged and the circumference of the hole decreases
   E the circumference of the disc remains unchanged and the circumference of the hole increases

2.10 When you sit in front of an open fire you are warmed by

   A convection, conduction and radiation
   B convection and conduction
   C radiation and convection
   D conduction only
   E radiation and conduction

2.11 A domestic supply of hot water is required for a short period of time in the morning and in the evening. The storage tank is covered with an insulating jacket to cut down heat loss and the water is heated by an immersion heater fitted with a thermostat. It is most economical to
   (i) heat the water from cold just before it is needed and switch off the heater in between

   OR

   (ii) leave the heater switched on all day so that the

thermostat can keep the temperature constant

A (ii) because more electrical energy would be needed to heat the water from cold

B (ii) because heat loss is minimised by the jacket

C (i) because the constant temperature would be more easily maintained for a short time than all day

D (i) because the rate at which heat is lost depends on the temperature difference between the water and the air temperature and is greatest when the tank is hottest

E (i) or (ii) depending how much water is used

2.12 The operation of a combined maximum and minimum thermometer (Six's design) is based on the thermal expansion of

A alcohol
B air
C mercury
D mercury and alcohol
E steel

2.13 Equal volumes of air, copper and water expand by different amounts when heated through the same temperature range, say 20°C to 30°C. If they are arranged in order of increasing expansion, starting with the one that expands least, the correct order is

A water, copper, air
B copper, water, air
C air, copper, water
D copper, air, water
E water, air, copper

2.14 In general, coastal resorts do not experience such large extremes of temperature as places inland. An important factor is the relatively high

A thermal conductivity of water

B   specific latent heat of vaporisation of water
C   specific heat capacity of water
D   thermal conductivity of the ground
E   specific heat capacity of the ground

2.15

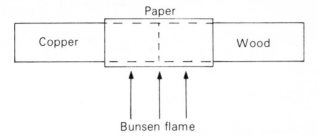

Bunsen flame

A rod is constructed half of copper and half of wood. A piece of paper is wrapped around the rod and heated over a Bunsen flame as shown above. The paper burns

A   in the middle since that part is directly over the flame
B   where it is covering the wood because wood is a better thermal conductor than copper
C   where it is covering the wood because wood is a poorer thermal conductor than copper
D   where it is covering the copper because wood is a better thermal conductor than copper
E   where it is covering the copper because wood is a poorer thermal conductor than copper

2.16   In a domestic hot water system, distribution of heat energy is mainly by

A   radiation
B   convection
C   conduction
D   conduction and radiation
E   radiation, convection and conduction

2.17 Some naphthalene is heated in a water bath until it has melted. The test tube containing the naphthalene is then placed in an empty beaker to cool. The temperatures of the naphthalene and the water are recorded every half minute as they cool. If cooling curves are plotted, which set of curves is most likely to be obtained? (W = water, N = naphthalene)

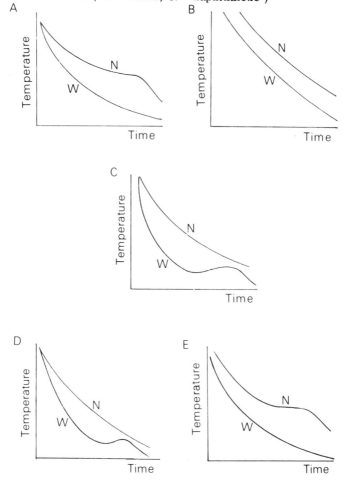

2.18 The specific latent heat of fusion of ice is 340 kJ kg^{-1}

When 10 kg of water freezes,

A absorbs 34 kJ of energy
B gives up 34 kJ of energy
C absorbs 3400 kJ of energy
D gives up 3400 kJ of energy
E either gives up or absorbs energy depending on the room temperature

2.19 Boiling water poured into a beaker is more likely to stress the beaker if it is made of glass than if it is made of copper because

A the glass will expand more than the copper
B the copper will expand more than the glass
C glass is a poorer thermal conductor than copper
D glass has a lower melting point than copper
E copper reflects more heat energy than glass does

2.20 When some of a liquid evaporates, the average speed of the molecules remaining will be

A increased because the more energetic molecules have left
B decreased because the more energetic molecules have left
C unchanged because all molecules have about the same speed
D increased because there are fewer molecules
E decreased because there are fewer molecules

2.21 A piece of copper of mass $0 \cdot 001$ kg is at a temperature of -34°C. It is placed in a beaker of water at 0°C. If the specific latent heat of fusion of ice is 340 kJkg^{-1} and the specific heat capacity of copper is $400\text{Jkg}^{-1}\text{K}^{-1}$, the mass of ice formed is

A $0 \cdot 004$ g
B $0 \cdot 04$ g
C $0 \cdot 4$ g
D $4 \cdot 0$ g
E $40 \cdot 0$ g

2.22

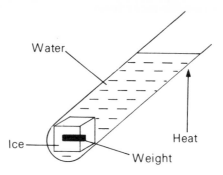

Water near the top of the test tube is heated and starts to boil before the ice starts to melt. This illustrates the fact that

A ice is a poor thermal conductor
B water is a good thermal conductor
C no convection occurs within the water
D water has a large specific latent heat of fusion
E water and ice have different specific heat capacities

2.23 Which one of the following is NOT a method of solving problems caused by the thermal expansion or contraction of solids

A the safety valve in a pressure cooker
B the gaps between lengths of railway lines
C bends in steam–carrying pipes
D the joints in large bridges
E installing overhead power lines in such a way that they sag in summer

2.24 A bimetallic strip bends when heated because the metals

A have different temperatures
B expand in opposite directions
C expand by different amounts
D become softer when heated and the weight causes the strip to bend

E   initially have different lengths and heating makes this difference even greater

2.25 Heat energy is transferred across a vacuum by

A   convection
B   radiation
C   radiation and conduction
D   convection and conduction
E   radiation and convection

2.26 A quantity of water is heated by an immersion heater from $10^{\circ}C$ to $40^{\circ}C$ in 10 minutes. If two such immersion heaters are used, they can raise the temperature of (neglect heat losses to the surroundings)

A   the same amount of water from $40^{\circ}C$ to $70^{\circ}C$ in 10 minutes
B   the same amount of water from $10^{\circ}C$ to $80^{\circ}C$ in 10 minutes
C   twice as much water from $40^{\circ}C$ to $70^{\circ}C$ in 20 minutes
D   half as much water from $10^{\circ}C$ to $40^{\circ}C$ in 5 minutes
E   half as much water from $10^{\circ}C$ to $70^{\circ}C$ in 5 minutes

2.27 Which one of the following is NOT an example of how poor thermal conductors are used to good effect in the home ?

A   a saucepan handle may be covered with plastic
B   an immersion heater tank is lagged
C   a refrigerator door has a solid metal handle
D   oven gloves are thickly padded
E   table mats are made of cork

2.28 When the bulb of a mercury-in-glass thermometer is heated, the FIRST thing that may be observed is that the mercury level drops because

A   the glass expands more than the mercury since glass has a greater expansion coefficient
B   the glass expands more than the mercury since

glass is a better thermal conductor

C the glass expands more than the mercury since glass has a larger specific heat capacity

D the heat reaches the glass first causing it to expand before the mercury

E air in the capillary tube expands and forces the mercury down

2.29 Equal masses of copper and aluminium are to be placed in a beaker of water at 0°C. The copper is initially at 50°C and the aluminium at 100°C. If the specific heat capacities of copper and aluminium are 400 $Jkg^{-1}K^{-1}$ and 880 $Jkg^{-1}K^{-1}$ respectively, the final water temperature will be

A between 0°C and 50°C

B between 50°C and 75°C

C between 75°C and 100°C

D 100°C

E dependent upon the mass of the water

2.30 Some campers boil water in a saucepan near the top of a mountain. If they boiled the same amount of water in the saucepan at the foot of the mountain, the air temperature being the same, then at the bottom,

A the water takes longer to reach the boiling point which is higher

B the water takes longer to reach the boiling point which is lower

C the water takes less time to reach the boiling point which is higher

D the water takes less time to reach the boiling point which is lower

E the water takes exactly the same time to reach the boiling point, but boils at a different temperature

OPTICS

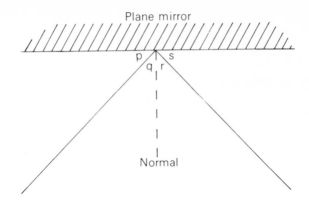

If the diagram represents the reflection of light from a plane mirror, then the angle of reflection is

A   p
B   q
C   r
D   s
E   impossible to determine from the information given

3.2   The apparent bending of a ruler when it is dipped into a beaker of water is due to

    A   reflection
    B   refraction
    C   diffraction
    D   dispersion
    E   interference

3.3   The optical centre of a lens must be

    A   the same point as the geometric centre
    B   equidistant from both faces of the lens
    C   the point through which light passes without deviating
    D   at the principal focus of the lens
    E   the centre of one of the faces of the lens

3.4 If you look out of a window through a ruffled net curtain you can see a pattern which is not that of the net. This pattern is caused by

A reflection
B refraction
C dispersion
D interference
E diffusion

3.5

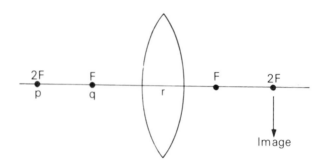

To produce an image in the position shown, the object would have to be placed

A between q and r
B at q
C between p and q
D at p
E to the left of p

3.6 In the following diagram, the angle of deviation is

A p
B q
C r
D s
E t

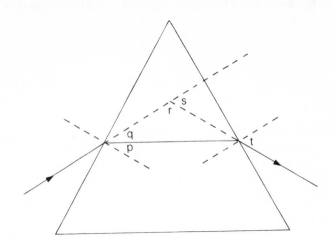

3.7 If an object is placed in front of a diverging lens, the image formed will be

A   real, on the same side of the lens as the object, and diminished

B   virtual, on the same side of the lens as the object, and diminished

C   virtual, on the same side of the lens as the object, and enlarged

D   real, on the opposite side of the lens to the object, and enlarged

E   real, on the opposite side of the lens to the object, and diminished

3.8 Snell's law concerning the passage of light from one medium to another states that

A   the sine of the angle of incidence equals the sine of the angle of refraction

B   the angle of incidence equals the angle of refraction

C   the ratio of the sine of the angle of incidence to the sine of the angle of refraction is constant

D   the ratio of the angle of incidence to the angle of refraction is constant

E   the incident, normal and reflected rays all lie in the same plane

3.9   If a beam of white light is passed through a prism, a spectrum is formed because rays of different colours and wavelengths are deviated by different amounts. Which of the following statements is true?

A   red has the longest wavelength and is deviated most

B   red has the longest wavelength and is deviated least

C   violet has the longest wavelength and is deviated most

D   violet has the longest wavelength and is deviated least

E   none of the above statements is true

3.10   When viewed in green light, a flag that is red and green will appear to be

A   black

B   white

C   green

D   green and black

E   green and yellow

3.11   1, 2 and 3 (in the diagram opposite) represent rays of light incident upon a face of a right prism. Before emerging from the prism, which ray (or rays) will experience total internal reflection?

A   1

B   2

C   3

D   1 and 3

E   none of them

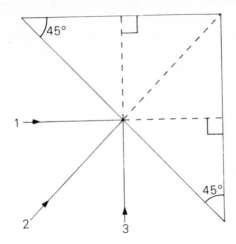

3.12 An object is placed at the principal focus of a diverging lens. The image formed is at a distance from the lens of

    A half a focal length on the same side as the object
    B half a focal length on the opposite side to the object
    C one focal length on the opposite side to the object
    D two focal lengths on the same side as the object
    E two focal lengths on the opposite side to the object

3.13 An opaque object is floating 3·5 cm above the bottom of a liquid, but appears to be floating 5·0 cm below the surface. If the liquid is 10·0 cm deep, its refractive index is

    A 0·70
    B 0·78
    C 1·19
    D 1·30
    E 1·43

3.14 Which one of the following statements about lenses is true ?

A   a diverging lens may produce a real or virtual image
B   a diverging lens always produces an inverted image
C   a diverging lens cannot produce a magnified image
D   a converging lens cannot produce a diminished image
E   a converging lens always produces an inverted image

3.15 White light produces a spectrum that is

A   a line spectrum
B   a band spectrum
C   an absorption spectrum
D   a continuous spectrum
E   a primary spectrum

3.16 A concave mirror of focal length 10 cm forms an erect image. If the image is twice the size of the object, then the object is at a distance from the mirror of

A   3·3 cm
B   5 cm
C   10 cm
D   15 cm
E   30 cm

3.17 Light from a sodium lamp passes through a single narrow slit and then through two close, parallel, narrow slits. If you look through the double slit towards the sodium lamp you will see

A   a continuous yellow band
B   a continuous black band
C   alternate black bands and yellow bands
D   the colours of the rainbow
E   a hazy yellow light

3.18 Fizeau attempted to measure the speed of light by using

A   lanterns on top of nearby mountains

B a laser
C a prism
D a rotating toothed wheel
E a diffraction grating

3.19 After the sun has sunk below the horizon, it can

    A be seen because light rays are refracted through the atmosphere towards the observer

    B not be seen because light rays are refracted through the atmosphere away from the observer

    C be seen because light rays are reflected off the atmosphere and into view

    D not be seen because light rays are reflected off the atmosphere out of sight

    E not be seen because light rays travel in straight lines

3.20 The colours seen on a soap bubble are a result of interference. When light strikes the bubble normally, the first thing that happens is that much of the light is

    A absorbed
    B reflected
    C refracted
    D diffracted
    E radiated

3.21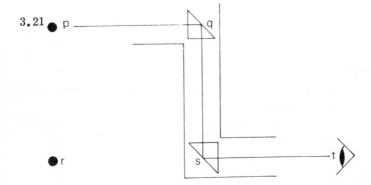

In the periscope shown above, the image will appear

at the point

A p
B q
C r
D s
E t

One light-year is equal to approximately

A 3 x 10^5 km

B 3 x 10^5 years

C 1 x 10^{13} km

D 1 x 10^{13} years

E 1 x 10^{20} s

3.23 Total internal reflection occurs when

A light passes from air to glass with an angle of incidence greater than the critical angle
B light passes from air to glass with an angle of incidence less than the critical angle
C light passes from glass to air with an angle of incidence greater than the critical angle
D light passes from glass to air with an angle of incidence less than the critical angle
E light passes from glass to air or from air to glass with an angle of incidence equal to 90°

3.24 The diagram on the opposite page best represents

A a compound microscope
B a telescope
C a magnifying glass
D a projector
E a camera

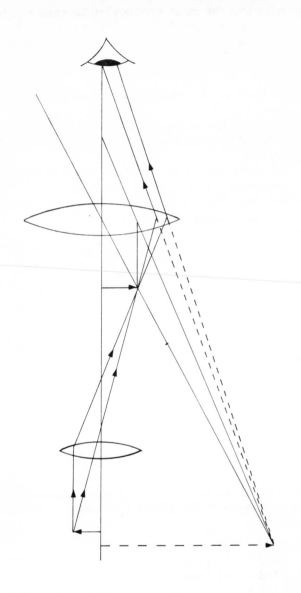

3.25 An eclipse of the moon occurs when the earth lies in a direct line between the sun and the moon. Such an eclipse does not occur every month, however, because

A the moon will be rotating about its own axis at that time

B the earth will be rotating about its own axis at that time

C the revolution of the moon about the earth is in a different plane to the revolution of the earth about the sun

D the earth takes more than one month to revolve around the sun

E the moon takes more than one month to revolve around the earth

3.26

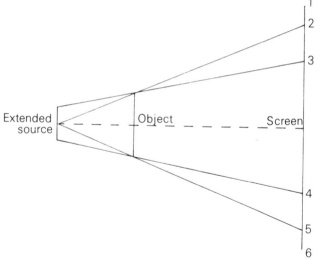

An object is illuminated by an extended source. The umbra consists of the region (or regions) between points

A 1 and 6

B 1 and 2, and 5 and 6

C 2 and 3, and 4 and 5

D 2 and 5

E 3 and 4

3.27 A ray of light shines onto a piece of matt white paper. The light will be mainly

A diffracted
B absorbed
C dispersed
D diffused
E scattered

3.28

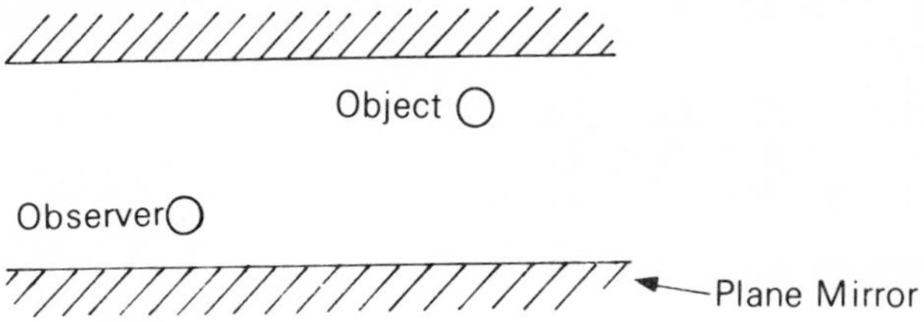

Object ◯

Observer◯

Plane Mirror

In the above diagram, the number of images that can be formed theoretically is

A 1
B 2
C 3
D 4
E infinite

3.29 If there is no parallax between two objects, then the two must be

A in the same place
B close to each other
C some way apart from each other
D virtual
E invisible

3.30 Rays of light parallel to the principal axis of a convex mirror will, after hitting the mirror,

A be reflected to one point
B be reflected as parallel rays
C converge to a point behind the mirror
D be reflected as though they had originated from roughly a point
E be refracted

WAVES AND SOUND

Questions 4.1 and 4.2 refer to the following diagram which represents the cross-section of a sound wave in air.

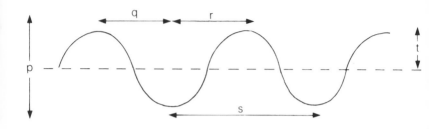

4.1 The wavelength is denoted by the distance

    A   p
    B   q
    C   r
    D   s
    E   t

4.2 The amplitude is denoted by the distance

    A   p
    B   q
    C   r
    D   s
    E   t

4.3 Which of the following phenomena can occur with sound waves ?

    A   interference only
    B   interference and diffraction
    C   interference and refraction
    D   diffraction and refraction

E   interference, diffraction and refraction

4.4 A rope, with its endpoints fixed, is shaken so that a stationary wave is produced. It there are four nodes, including the two at the fixed ends, the number of anti-nodes is

A   1
B   2
C   3
D   4
E   5

4.5 The vibration with the greatest wavelength that can be produced on a guitar is

A   the decibel
B   the fundamental
C   the second harmonic
D   an overtone
E   an octave

4.6 A whistle is blown and an echo is heard from a distant cliff three seconds later. If the speed of sound in air is $0 \cdot 33 \text{ kms}^{-1}$, how far away is the cliff?

A $0 \cdot 05$ km
B $0 \cdot 1$ km
C $0 \cdot 5$ km
D   1 km
E   3 km

4.7 The frequency of a plucked string

A   increases when the length is increased and increases when the tension is increased
B   increases when the length is increased and decreases when the tension is increased
C   decreases when the length is increased and increases when the tension is increased
D   decreases when the length is increased and decreases when the tension is increased
E   depends only upon the tension in the string

4.8 During a telephone conversation, the caller's voice often sounds different from when you are talking to him face to face. The most likely explanation is that

A only a limited frequency range is being transmitted
B the overall energy is reduced because of the distance between you and the caller
C the 'receiver' into which the caller speaks does not receive all the sound because of diffraction
D sound, being a wave motion, exhibits interference
E there are often faulty connections and crossed lines in the telephone system

4.9 Which one of the following suggests that a material may not always be essential for the propagation of a wave ?

A sound travels through the air
B sound cannot travel through a vacuum
C light reaches us from the sun
D a piece of wood floating in water is not carried along by a water wave
E a tuning fork may be vibrated in an evacuated container

4.10 A stationary wave is formed as a result of

A diffraction
B interference
C absorption
D resonance
E refraction

4.11 On a still night, sound may be heard over longer distances than in the daytime because

A the air near the earth is warmer than higher up and the sound wavefronts are refracted towards the earth
B the air near the earth is cooler than higher up and the sound wavefronts are refracted away from the earth

C   the air near the earth is cooler than higher up
    and the sound wavefronts are refracted towards
    the earth

D   the layers of air near the earth are cooler than
    higher up and the sound travels more quickly

E   the layers of air near the earth are warmer than
    higher up and the sound travels more slowly

4.12 Middle C played on a piano always sounds different
      from when it is played on a violin because of a
      difference in

A   pitch
B   frequency
C   amplitude
D   beats
E   harmonics

4.13 The note that is one octave above the note with
      frequency 100 Hz has a frequency of

A   100 Hz
B   108 Hz
C   200 Hz
D   800 Hz
E   1 kHz

4.14 If radio waves travel at a speed of 3×10^5 km s^{-1}, then
      the wavelength of a radio wave of frequency 1·5 MHz
      is

A   0·20 m
B   5·0 m
C   200 m
D   450 m
E   500 m

4.15 When you blow across the 'mouth' of a test tube
      containing some water, a certain sound is produced.
      If you repeat this process after more water has been

added to the test tube, the pitch of the sound will be

A higher because the length of the water column is greater
B higher because the length of the air column is less
C the same as before
D lower because the length of the water column is greater
E lower because the length of the air column is less

4.16 During a thunderstorm, the flash of lightning is seen before the thunder is heard. This is because

A light travels faster than sound
B sound travels faster than light
C the lightning is produced before the thunder
D the thunder is produced before the lightning
E the eye is more sensitive than the ear

4.17

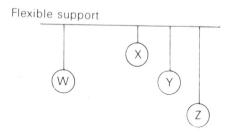

Flexible support

The brass pendulum W is set swinging from the flexible support and forced vibrations cause the three wooden bobs X, Y and Z to swing as well. Which of the following statements about the bobs X, Y and Z is correct?

A X swings with the greatest amplitude

B   Y swings with the greatest amplitude

C   Z swings with the greatest amplitude

D   X, Y and Z all swing with the same amplitude

E   X and Z swing with a greater amplitude than Y

4.18   Compared with the speed of sound in a gas at a given temperature and pressure, the average speed of the gas molecules is

A   considerably smaller

B   a little smaller

C   the same

D   greater

E   smaller or greater depending on which gas is being considered

4.19   A detonator is fired from a ship at sea.   The sound is reflected from the sea bed, the time between the sound being produced and the echo being received is one third of a second. Assuming the speed of sound in sea water is a constant $1 \cdot 50 \text{ km s}^{-1}$, how deep is the sea at this point?

A $0 \cdot 125$ km

B $0 \cdot 250$ km

C $0 \cdot 500$ km

D $4 \cdot 50$ km

E $22 \cdot 5$ km

4.20   When a band is heard a long way away, the piccolo and the tuba do not seem to get out of step with each other because

A   they produce notes of the same frequency

B   waves of different frequency travel at the same speed

C   the pitch of the two sounds is the same

D   both sets of waves travel directly without diffraction

E   by the time the ear has received the sounds, they are in step again

4.21 A parabolic barrier is set up in a ripple tank. Circular waves are generated at a point in front of the barrier. Which of the following best describes what happens to these waves when they hit the barrier?

A they are reflected to a single point
B they are reflected in such a way that they diverge
C they are reflected as parallel straight waves
D they are reflected in random directions
E the way they are reflected depends on the precise location of the source

4.22 All forms of electromagnetic waves travelling in a vacuum have the same

A amplitude
B frequency
C wavelength
D speed
E period

4.23 A single pulse is sent along a rope. What is actually transferred from one end to the other is

A energy
B rope
C velocity
D the medium
E the source

4.24 A tuning fork has a frequency of 272 Hz. The velocity of sound in air is 340 ms^{-1}. The wavelength of the sound the tuning fork produces is

A 6·8 m
B 6·12 m
C 3·4 m
D 1·25 m
E 0·8 m

4.25 For a transverse wave of constant velocity, which of

the following depend upon each other?

A amplitude and frequency
B amplitude and wavelength
C wavelength and frequency
D amplitude, frequency and wavelength
E none of the above

4.26 The disc of a siren rotates 600 times each minute. If the disc has 25 equally spaced holes in it, then the frequency of the note emitted is

A 4·2 Hz
B 10 Hz
C 250 Hz
D 1440 Hz
E 15 kHz

4.27 A transverse wave vibrating in one direction only is said to be

A polarized
B diffracted
C resonating
D stationary or standing
E visible

4.28 A sound wave travelling through air establishes regions in which the pressure is slightly less than normal. These regions are known as

A condensations
B compressions
C rarefactions
D vacua
E interference

4.29

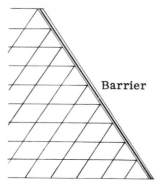

Barrier

The diagram above shows periodic straight waves in a ripple tank in which there is a barrier. The phenomenon represented is

A   reflection
B   refraction
C   diffraction
D   interference
E   dispersion

4.30

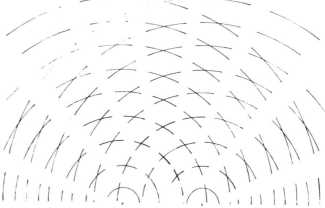

The pattern of water waves in a ripple tank shown above is best obtained by using

A   two vibrating point sources
B   two vibrating point sources and a straight barrier
C   straight waves and a barrier with a single slit
D   a single vibrating point source
E   straight waves and a curved barrier

5.1 The effect of heating a magnet is to

A   strengthen it
B   weaken it
C   reverse the polarity
D   leave it unaffected
E   do something else

5.2

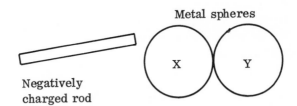

Metal spheres

X   Y

Negatively
charged rod

Two metal spheres, each supported on an insulating
stand, are placed in contact with each other. A
negatively charged rod is brought close to sphere X,
but is not touching it.   If sphere Y is now moved away
from sphere X and the rod removed, the two spheres
will be charged as follows:

A   both negative
B   both positive
C   X: positive, Y: negative
D   X: negative, Y: positive
E   both uncharged since the rod did n   .ctually touch
    X or Y

5.3 Arranged in order of increasing e.m.f., the dry cell,
     lead-acid accumulator and Weston cell would be listed
     as

A   dry cell, accumulator, Weston
B   dry cell, Weston, accumulator
C   Weston, accumulator, dry cell

D Weston, dry cell, accumulator

E accumulator, Weston, dry cell

5.4 Two long straight wires are parallel and close to each other, but not touching. A steady current is passed through each, the two currents being in the same direction.

    A the wires do not move since the currents are not varying

    B both wires revolve about one another

    C the wires attract each other

    D the wires repel each other

    E the wires oscillate together

5.5

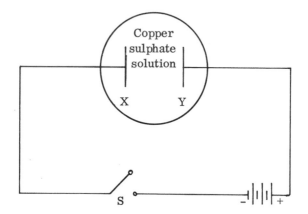

X and Y are two copper plates suspended in a copper sulphate solution and connected as shown. Switch S is closed and the apparatus allowed to operate for half an hour. When the two plates are weighed, it is found that their masses have changed:

    A X: increased, Y: decreased

    B X: increased, Y: increased

    C X: decreased, Y: increased

    D X: decreased, Y: increased

    E X: increased, Y: unchanged

5.6 Which one of the following statements about a wire and a horse-shoe magnet is false ?

A no e.m.f. is induced when the wire is stationary and some distance from the magnet

B no e.m.f. is induced when the wire is stationary within the magnetic field

C an e.m.f. is induced when the wire is moved within and parallel to the magnetic field

D an e.m.f. is induced when the wire is moved across the magnetic field

E no e.m.f. is induced when the wire is moved while it is some distance from the magnet

5.7 Three metal slabs, X, Y and Z, identical in shape and size, are suspected of being magnets. Tests are carried out and it is found that there is attraction between poles 2 and 5, and between poles 2 and 4, but poles 2 and 3 repel each other. Without making further tests, you could correctly conclude that

A all three slabs are magnets

B X and Y are magnets, but Z is not

C poles 1 and 4 would attract each other

D Z cannot possibly be a magnet

E if Z is a magnet, then poles 1 and 5 would repel each other

| 1 | X | 2 | 3 | Y | 4 | 5 | Z | 6 |

5.8

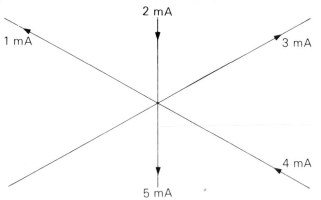

2 mA

1 mA

3 mA

4 mA

5 mA

Various currents flow in and out of a junction in a complex circuit. The current I will be

A   6 mA into the junction
B   6 mA out of the junction
C   3 mA into the junction
D   3 mA out of the junction
E   0

5.9   A normal domestic electricity supply is an alternating current whose average value is

A   zero
B   half the peak value
C   the peak value multiplied by $\sqrt{2}$
D   the peak value divided by $\sqrt{2}$
E   dependent upon the voltage

5.10   An a.c. supply may be used directly for all but which one of the following?

A   heating
B   lighting
C   transforming voltage
D   electroplating
E   blowing a fuse

5.11 A commutator is used in

    A a transformer
    B an induction coil
    C a d.c. motor
    D a microphone
    E an electric bell

5.12 Which one of these statements about magnetic induction is correct?

    A magnetism may be induced in any material
    B the material must be in contact with the magnet before magnetism is induced
    C the magnetism is permanent
    D magnetic induction can explain why unmagnetised objects may be attracted by a magnet
    E an electromagnet cannot be used to induce magnetism

5.13 A voltameter is used for

    A studying electrolysis
    B measuring resistance
    C measuring potential difference and current
    D measuring temperature
    E studying the flow of electricity in a gas

5.14 When a 48 ohm resistor is connected across the terminals of a 6 volt battery of negligible internal resistance, the power developed in the resistor is

    A 0·125 W
    B 0·75 W
    C 1·3 W
    D 8·0 W
    E 10·7 W

5.15 The dome of a Van de Graaff generator is smooth and spherical because

    A a sphere is the shape that permits maximum volume

B heat energy produced is easily radiated

C charged particles in the air are easily reflected

D the shape reduces the likelihood of electrostatic breakdown

E a symmetrically shaped dome is easier to charge

5.16 A boy purchases a bar magnet whose poles are marked N and S in the usual manner. If he suspends the magnet in a horizontal position so that it is free to rotate, the pole marked N would point towards

A the South Magnetic Pole

B the South Geographic Pole

C the North Magnetic Pole

D the North Geographic Pole

E a point somewhere between a magnetic pole and a geographic pole

5.17

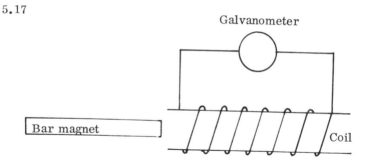

Maximum deflection of the galvanometer needle occurs when

A the magnet is pushed into the coil

B the magnet is rotated inside the coil

C the magnet is stationary within the coil

D the number of turns of wire in the coil is reduced

E a copper bar is used in place of the bar magnet

5.18 A domestic electricity meter measures in units of

A volt
B kilowatt
C kilowatt-hour
D ampere
E ampere-hour

5.19 If an ammeter is inadvertently connected directly across a voltage supply, the most likely thing that will happen is

A it will measure voltage, but only approximately
B nothing
C it will get hot and perhaps burn out
D dependent on whether the ammeter is a moving coil or moving iron type
E dependent on whether the voltage supply is a.c. or d.c.

5.20 Which one of the following equations concerning a capacitor is correct?

A capacitance = charge x potential difference
B charge = capacitance x potential difference
C potential difference = charge x capacitance
D potential difference = $\dfrac{\text{capacitance}}{\text{charge}}$
E capacitance = $\dfrac{\text{potential difference}}{\text{charge}}$

5.21 A famous scientist once described a piece of equipment as 'only an assemblage of a number of good conductors of different sorts arranged in a certain way. 30, 40, 60 pieces or more of copper ... each in contact with a piece of tin ... and an equal number of layers of water or some other liquid ...' The scientist was referring to a

A resistor
B battery
C voltmeter

D capacitor

E transformer

5.22 If the flashgun of a camera operates for a millisecond and during this time 0·05 coulombs of charge flow, then the current will be

A 5×10^{-8}A

B 5×10^{-5}A

C 0·02 A

D 50 A

E 5 kA

5.23 One reason why the efficiency of a transformer is usually high is

A only small currents are used

B a.c. is used

C there are no moving parts

D the output voltage is greater than the input voltage

E the primary and secondary coils are some distance apart

Questions 5.24 and 5.25 refer to the following diagram

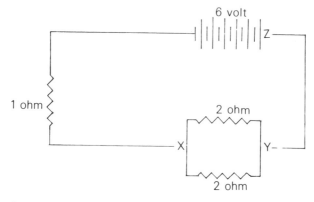

5.24 Assuming the internal resistance of the battery is negligible, a voltmeter connected between points X

and Y should read

A 0 V
B 1·2 V
C 3·0 V
D 4·8 V
E 6·0 V

5.25 Assuming the internal resistance of the battery is negligible, a voltmeter connected between points X and Z should read

A 0 V
B 1·2 V
C 3·0 V
D 4·8 V
E 6·0 V

5.26 A small electric motor is connected to a battery, but fails to start because of friction. The motor should immediately be started by hand or disconnected because otherwise

A the potential difference across the motor will build up
B the motor will stick permanently
C the current will be too high
D no current will flow and the battery will be damaged
E the resistance of the coil of the motor will fall

5.27 Household electrical appliances are not usually connected in series because

A switching off one appliance would switch off the rest
B interference to television and radio is much greater
C a fuse would blow as soon as one appliance was used
D power consumption would be very much greater
E the potential drop across each appliance would be much higher and the appliances would be damaged

5.28 Three identical dry cells are connected in parallel. Compared with the e.m.f. and internal resistance of one cell, the total e.m.f. and the total internal resistance will be as follows:

	Total e.m.f	Total internal resistance
A	higher	higher
B	higher	lower
C	the same	higher
D	the same	lower
E	the same	the same

5.29 One metre of copper wire and one metre of iron wire are connected together in series with a battery. If the two wires have equal diameters, which one of the following observations is true?

A   the resistance of the copper is lower than that of the iron; more current flows in the copper wire which gets hotter than the iron wire

B   the resistance of the iron is lower than that of the copper; more current flows in the iron wire which gets hotter than the copper wire

C   the same current flows in each wire and they both get equally hot

D   the same current flows in each wire, but the iron wire gets hotter than the copper wire

E   the same current flows in each wire, but the copper wire gets hotter than the iron wire

5.30

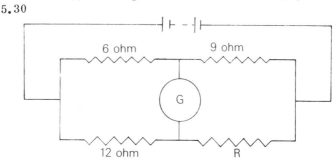

When the four resistors are connected as shown, it is

found that the current flowing through the galvanometer G is zero. The resistance of R is

A 3 ohm
B 4·5 ohm
C 15 ohm
D 18 ohm
E impossible to determine without knowing the e.m.f of the battery

ATOMIC AND NUCLEAR PHYSICS

6.1 Which one of the following is not used for the detection of radiation?

    A   photographic emulsion
    B   cyclotron
    C   spinthariscope
    D   cloud chamber
    E   Geiger-Müller tube

6.2 The helium isotope $^{3}_{2}\text{He}$

    A   has three protons and two neutrons
    B   has two protons and one neutron
    C   has two protons and three neutrons
    D   has one proton and two neutrons
    E   cannot exist because it has the same mass number as tritium

6.3 Gamma rays may be absorbed by

    A   a piece of paper
    B   a thin sheet of metal
    C   a cloud of steam
    D   a block of lead
    E   a column of air

6.4 A magnetic field may be used to deflect

    A   gamma rays
    B   alpha and gamma rays
    C   alpha and beta rays
    D   gamma and beta rays
    E   alpha, beta and gamma rays

6.5 The half-life of a radioactive isotope may be increased by

    A   increasing the mass present
    B   raising the temperature

C   increasing the pressure on the element

D   increasing the illumination

E   none of the above methods

6.6   The discovery of radioactivity is credited to

A   Becquerel

B   Pierre Curie

C   Marie Curie

D   Roentgen

E   Wilson

6.7   Radon-220 will eventually decay to bismuth-212 as

$$^{220}_{86}Rn \longrightarrow {}^{216}_{84}Po + {}^{4}_{2}He \quad \text{Half life:} \quad 55 \text{ seconds}$$

$$^{216}_{84}Po \longrightarrow {}^{212}_{82}Pb + {}^{4}_{2}He \quad \text{Half life: } 0.16 \text{ seconds}$$

$$^{212}_{82}Pb \longrightarrow {}^{212}_{83}Bi + {}^{0}_{-1}e \quad \text{Half life: } 10.6 \text{ hours}$$

If a certain mass of radon-220 is allowed to decay in a sealed container, after five minutes the element with the greatest mass will be

A   radon

B   polonium

C   helium

D   lead

E   bismuth

6.8   Which one of the following statements is true ?

A   the mass of a proton is about the same as that of an electron

B   the mass of an electron is about the same as that of a neutron

C   the mass of a proton is much greater than that of an electron

D   the mass of a neutron is much greater than that of a proton

E the masses of a proton, a neutron and an electron are all about the same

6.9

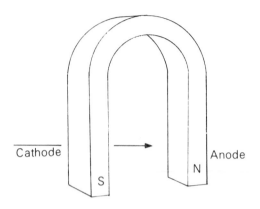

A stream of cathode rays passes from left to right between the poles of a powerful magnet. The cathode rays will be

A undeflected
B deflected downwards
C deflected upwards
D deflected towards the S pole
E deflected towards the N pole

6.10 An electron carries a charge of magnitude $1 \cdot 6 \times 10^{-19}$ coulomb. The magnitude of the charge on an alpha particle is

A $1 \cdot 6 \times 10^{-19}$ coulomb

B $3 \cdot 2 \times 10^{-19}$ coulomb

C $3 \cdot 2 \times 10^{-38}$ coulomb

D $6 \cdot 4 \times 10^{-19}$ coulomb

E $6 \cdot 4 \times 10^{-76}$ coulomb

6.11 X-rays can cause a gold leaf electroscope to discharge

    A   when the electroscope is positively charged only
    B   when the electroscope is negatively charged only
    C   when the electroscope is charged positively or negatively
    D   never
    E   only when the electroscope has been charged by X-rays

6.12 Millikan's oil drop experiment was designed to calculate the

    A   charge on an electron
    B   ratio of charge to mass of an electron
    C   charge on an ion
    D   ratio of charge to mass of an ion
    E   number of electrons in an oil drop

6.13 When an old television set is operated, it is found that X-rays are being produced. This suggests the presence of

    A   an air leak into the tube
    B   a very high voltage
    C   irregular coating of the screen
    D   a loose connection to the power supply
    E   overheating of the filament

6.14 In a so-called neon light, electricity is conducted by means of

    A   gamma rays
    B   incandescence
    C   ions and cathode rays
    D   neutral atoms
    E   fluorescence

6.15 Isotopes of the same element have

    A   different numbers of electrons
    B   the same number of neutrons

C different numbers of protons

D different mass numbers

E different atomic numbers

6.16 A diode valve may be used to provide a constant current even when the potential difference is varying because

A Ohm's law does not apply to a gaseous conductor and so the current is not affected at all by the voltage

B there is no grid present to cause variation in the current

C the maximum steady current, the saturation current produced by the cathode of the valve depends only on the steady temperature of the filament

D the current value depends only upon the size and spacing of the electrodes and therefore is constant

E after the valve has been operating for a short time, the positive potential applied to the anode drops to a low value

6.17 In a nuclear reactor the uranium-235 nucleus may be split by bombarding it with a neutron. Boron steel rods are often used to control the reaction because

A boron is chemically inert

B boron absorbs neutrons

C boron reflects neutrons

D boron and uranium-235 repel each other

E boron is a good thermal insulator

6.18 The mass number of an atom is

A the number of protons in the nucleus

B the number of electrons in the atom

C the number of nucleons in the nucleus

D exactly the same as the atomic weight

E the number of neutrons in the nucleus

6.19 Which one of these statements about cathode rays is untrue ?

A   they travel in straight lines
B   X-rays may be emitted when cathode rays hit a solid target
C   they are more penetrating than gamma rays
D   they can be deflected by means of a magnetic field
E   they carry a negative charge

6.20 The charge on an electron is about

A $1 \cdot 6 \times 10^{19}$ coulomb

B $1 \cdot 6 \times 10^{29}$ coulomb

C $1 \cdot 6 \times 10^{-19}$ coulomb

D $1 \cdot 6 \times 10^{-29}$ coulomb
E   none of these

6.21 Which one of the following statements is true ?

A   beta particles travel faster than alpha particles
B   beta particles travel faster than light
C   gamma rays travel more slowly than light
D   X-rays travel faster than gamma rays
E   X-rays can be made to travel faster by applying a very high voltage

6.22 The critical mass of uranium-235 is

A   the value of m in the equation $E = m c^2$
B   the maximum mass that can react at one time
C   the minimum mass that can react at one time
D   the minimum mass that is necessary for a chain reaction to occur
E   the maximum amount of energy that can be derived from 1 kg of uranium-235

6.23 A wire is heated to a high temperature and electrons are given off. This is known as

A photoelectricity
B fluorescence
C phosphorescence
D thermionic emission
E scintillation

6.24 Beta rays consist of

A protons and neutrons
B electrons
C protons
D electromagnetic waves
E helium nuclei

6.25 If a graph is made of the mass of a piece of radioactive element plotted against time, the graph will be

A a straight horizontal line
B a straight line with a negative slope
C a curve whose slope gets more negative as time increases
D a curve whose slope gets less negative as time increases
E a curve of irregular shape

6.26 The end product of a chain of radioactive decays is usually some form of

A helium
B hydrogen
C lead
D thorium
E uranium

6.27 What is missing from this equation?

$$_2^4\text{He} \;+\; _4^9\text{Be} \;\longrightarrow\; _6^{12}\text{C} \;+\; ?$$

A a proton

B   an electron
C   a neutron
D   hydrogen
E   lead

6.28   A ratemeter connected to a Geiger-Müller tube gives a count rate of 2000 per minute when a radioactive source is placed nearby. If the half life of the source is 30 minutes, then after 2 hours the count rate would be

A   1000 per minute
B   500 per minute
C   250 per minute
D   125 per minute
E   62·5 per minute

6.29   Rutherford's experiment in 1911, in which alpha particles were directed against a thin gold sheet, led directly to the concept of

A   the electron
B   the neutron
C   the proton
D   isotopes
E   the nucleus

6.30   Given that the following forms of helium exist, how many isotopes of helium exist?

$$_{2}^{3}He, \quad _{2}^{4}He, \quad _{2}^{5}He, \quad _{2}^{6}He$$

A   0
B   1
C   2
D   3
E   4

TEST I

7.1 A load is to. be raised by means of either a single fixed pulley system or a single movable pulley system. The mechanical advantages would be

Fixed Pulley	Movable Pulley
A greater than 1	greater than 1
B less than 1	less than 1
C greater than 1	less than 1
D less than 1	greater than 1
E exactly 1	exactly 1

7.2 A convex mirror forms a virtual image somewhere between the mirror and the principal focus. To do this, the object must be placed at a distance from the mirror of

A zero
B less than one focal length
C one focal length
D between one and two focal lengths
E any distance

7.3 A gun of mass 2·5 kg fires a bullet of mass 0·10 kg. If the initial speed of the bullet is about 600m s^{-1}, then the gun will recoil with a speed of about

A $2 \cdot 0 \times 10^{-4} \text{ m s}^{-1}$
B $4 \cdot 0 \times 10^{-4} \text{ m s}^{-1}$
C $0 \cdot 25 \text{ m s}^{-1}$
D 24 m s^{-1}
E 600 m s^{-1}

7.4 The kinetic energy of a car of mass 1000 kg travelling at a speed of 30 m s^{-1} is

A 30 kJ
B 450 kJ
C 900 kJ
D 3750 kJ

E $2 \cdot 25 \times 10^5$ kJ

7.5 The colour that is complementary to red is

A cyan
B magenta
C blue
D green
E yellow

7.6 There are 500 turns of wire in the primary coil of a transformer and 50 turns in the secondary coil. If the input voltage is 250 volt, then, neglecting heat energy losses, the output voltage will be

A   25 V
B   50 V
C   100 V
D   500 V
E 2500 V

7.7

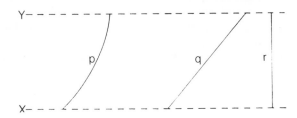

A heavy weight is to be lifted from level X to level Y. Which direction of lifting requires the least amount of work?

A p
B q
C r
D no work is required for p, q, or r
E the same amount of work is required for p, q and r

7.8 If you look closely at a bright metal sphere, such as a Christmas tree ornament, you will notice that the image of yourself is

A the right way up and larger than you
B the right way up and the same size as you
C the right way up and smaller than you
D upside down and larger than you
E upside down and smaller than you

7.9 Doubling the absolute temperature of a gas and halving its pressure will make the volume

A four times as large
B twice as large
C unchanged
D half as large
E one fourth as large

7.10 Ideally, ice skates should be very sharp because this has the effect of

A increasing the pressure on the ice and raising its melting point
B increasing the pressure on the ice and lowering its melting point
C decreasing the pressure on the ice and raising its melting point
D decreasing the pressure on the ice and lowering its melting point
E none of the above

7.11 A fresh egg is placed at the bottom of a jar of water. The best way to make the egg float is to

A add more water to the jar
B raise the water temperature by about 10°C
C lower the water temperature by about 10°C
D dissolve some salt in the water
E lower the water level until it is the same as the uppermost part of the egg

7.12 Frequency may be measured in units of

A second
B Hertz
C metre
D metre/second
E metre⁻1

7.13 A man is able to distinguish two sounds if they are at least 0·1 second apart. If sound travels at 0·33 kms⁻¹, how close to a wall can the man stand and hear the echo of his voice?

A 330 m
B 660 m
C 165 m
D 33 m
E 16·5 m

7.14 For light passing from glass to air, the critical angle is

A 0^0
B between 0^0 and 30^0
C between 30^0 and 60^0
D between 60^0 and 90^0
E 90^0

7.15 Three suggested ways in which a guitarist could produce a note of higher pitch are:

(i) shorten the length of string
(ii) reduce the tension in the string
(iii) use a lighter string

Which would have the desired effect?

A only one of these suggestions
B (i) and (ii)
C (i) and (iii)
D (ii) and (iii)
E all three suggestions

7.16 A sports car overtakes your saloon car along the motorway. As it gets further away from you, the pitch of the exhaust sound

A increases according to the Doppler effect
B decreases according to the Doppler effect
C is constant according to the Doppler effect
D is constant since the Doppler effect does not apply to two bodies moving in the same direction
E increases or decreases, depending on whether the passing car is accelerating or travelling at constant speed

7.17 The altitude of a plane can be most easily found by means of a

A gyroscope
B hydrometer
C barometer
D commutator
E spectroscope

7.18 The cap of a gold leaf electroscope is touched with an object but this has no effect at all. You can only conclude that

A the object must be a conductor
B the object must be an insulator
C the electroscope was not charged originally
D the original charge on the electroscope was of the same sign as the charge on the object
E the original charge on the electroscope was of the opposite sign to the charge on the object

7.19 The wave that results when a stone is dropped into a pond is a

A stationary wave
B shock wave
C light wave
D progressive wave
E microwave

7.20 Valves in a radio set are often surrounded by metal covers. The purpose of these is to

A   prevent breakage of the glass tubes
B   prevent other components in the set from getting too hot
C   maintain the operating temperatures of the valves at a constant level
D   shield the valve from external alternating fields
E   reduce the chances of short circuits occurring between neighbouring valves

7.21 A light bulb is marked 60 W, 240 V, and is used on a supply of 240 V. The current that is drawn is (in amps)

A   2·5
B   0·25
C   0·4
D   4
E   6

7.22 Two insulated beakers each contain equal masses of water at 22°C. A third beaker, identical to the first two, contains a liquid whose specific heat capacity is half that of water and whose mass is equal to the total mass of water in the other two beakers. The initial temperature of this liquid is 100°C. If the contents of all three beakers are mixed together and there is only negligible heat energy transfer to the surrounding air, the final temperature of the mixture will be

A   37·6°C
B   48·0°C
C   59·3°C
D   61·0°C
E   70·0°C

7.23 The cost of operating a 600 W hairdryer for 20 minutes, if electricity costs $2\frac{1}{2}$ p per unit (kW-hour), will be

A   1·0p

B   0·75p
C   0·5p
D   0·25p
E   less than 0·25p

7.24   A small object is to be placed somewhere along the principal axis of a concave mirror so that the image formed coincides with the object.   The object should be placed

   A   at the principal focus
   B   at the centre of curvature
   C   at the pole of the mirror
   D   between the pole and the principal focus
   E   more than two focal lengths away from the mirror

7.25   The temperature of some weighed mercury is noted and the mercury is poured into a long glass tube. After this has been carefully sealed at both ends, it is inverted one hundred times.   If the temperature of the mercury is again recorded and the length of the tube through which the mercury fell each time is measured, you could calculate the

   A   specific latent heat of mercury
   B   specific heat capacity of mercury
   C   the kinetic energy of the mercury molecules
   D   atmospheric pressure
   E   coefficient of thermal expansion of mercury

7.26   An electricity supply is described as 240 V, a.c.   In fact, the potential difference will vary between about

   A   0 and 240 volt
   B   −240 and 240 volt
   C   0 and 340 volt
   D   −340 and 340 volt
   E   228 and 252 volt

7.27   Gamma rays are rarely detected in a cloud chamber because

A   only electromagnetic radiation can be detected in
    a cloud chamber
B   gamma rays often fail to interact in the chamber
    and so produce detectable ions
C   gamma rays are invisible to the naked eye
D   gamma rays are not very penetrating
E   gamma rays are not charged particles

7.28   The resistance of a certain piece of wire is 9 ohm.
       The resistance of a second piece, identical to the first
       except that its diameter is three times as large, will
       be

A   1 ohm
B   3 ohm
C   9 ohm
D   27 ohm
E   81 ohm

7.29

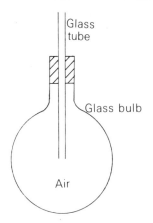

The glass bulb is held tightly in both hands.   At
first

A   the air is compressed by the pressure of the hands
    and is forced out along the glass tube
B   the temperature of the glass bulb increases, the
    bulb expands and more air is drawn into the bulb
C   the bulb is compressed by the pressure of the
    hands and air is forced out along the glass tube
D   nothing happens since the air is invisible
E   nothing happens because air and glass expand
    identically

7.30 An important difference between a microscope and a telescope is that

A   the object lens of the telescope produces a real image, but the object lens of the microscope produces a virtual image

B   the object lens of the telescope produces a virtual image, but the object lens of the microscope produces a real image

C   the focal length of the telescope object lens is much greater than that of the microscope object lens

D   the focal length of the microscope object lens is much greater than that of the telescope lens

E   the final image in a telescope is always inverted

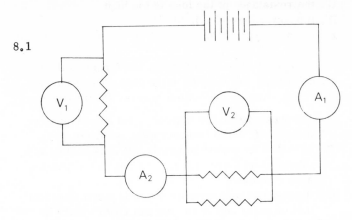

8.1

Three identical resistors are connected with voltmeters V_1 and V_2, and ammeters A_1 and A_2 as shown. Which one of the following statements is true?

A A_1 and A_2 will show different readings

B A_1 will read higher than A_2

C V_1 and V_2 will read the same

D V_2 will read less than V_1

E the e.m.f. of the battery will be less than the sum of the readings on V_1 and V_2

8.2 Hare's apparatus is used for determining

A atmospheric pressure
B density
C temperature
D pressure in a liquid
E surface tension

8.3 In a normal household electricity supply a fuse blows when

A the voltage drop across it is too low
B the current flowing through it is too high

C   the resistance of the fuse is too high

D   the power expended is too low

E   too much charge has built up on one side of the fuse

8.4   When a pond freezes in winter, ice forms on the top first because the surface is in contact with the cold air. The water at the bottom may not freeze because

A   the pressure at the bottom is very much greater and therefore the freezing point is higher

B   as water is cooled to nearly $0^{\circ}C$, its volume increases and there is less room for this expansion at the bottom

C   the specific heat capacity of ice is much less than that of water and so the ice can absorb more heat energy

D   the thermal conductivity of water is much lower at lower temperatures and thus the water at the bottom is not cooled so much

E   water has a maximum density at $4^{\circ}C$ and water at this temperature cannot rise to be cooled further

8.5   An achromatic lens is usually made of

A   two pieces of plane glass

B   two converging lenses

C   two diverging lenses

D   one lens and a thin piece of plane glass

E   two lenses of different glass and focal lengths

8.6   The equation

$$^{4}_{2}He \ + \ ^{14}_{7}N \ \longrightarrow \ ^{17}_{8}O \ + \ ^{1}_{1}H$$

tells us that when the nitrogen nucleus is struck by

A   an alpha particle, an electron is emitted

B   an alpha particle, a proton is emitted

C   an alpha particle, a beta particle is emitted

D   a proton, a beta particle is emitted

E   a proton, an electron is emitted

8.7 The image formed by a plane mirror is

A virtual, behind the mirror and the same size as the object

B real, behind the mirror and the same size as the object

C virtual, at the surface of the mirror and the same size as the object

D real, at the surface of the mirror and enlarged

E virtual, behind the mirror and enlarged

8.8 The capacitance of a capacitor is measured in units of

A coulomb
B farad
C joule
D watt
E coulomb-watt

8.9 A battery has an e.m.f. of 12 volt and an internal resistance of 0.5 ohm. When an external 5.5 ohm resistor is connected across the terminals of the battery, the potential difference between the terminals will be

A 12 V
B 11.5 V
C 11 V
D 2 V
E 1 V

8.10 A man who is 1.40 m tall stands in front of a mirror and can just see himself from head to toe. Assuming that his eyes are 0.14 m below the top of his head, the height of the mirror is

A 1.40 m
B 1.26 m
C 0.70 m
D 0.63 m
E impossible to determine without knowing how far the man is from the mirror

8.11 A converging lens of focal length 10 cm is to be used to produce a real image that is five times as large as the object. The object must be placed at a distance from the lens of

A   2 cm
B   8 cm
C   10 cm
D   12 cm
E   60 cm

8.12 A 10 cm cube floats in water with a height of 4 cm remaining above the surface. The density of the material from which the cube is made is

A $0 \cdot 6$ g cm^{-3}
B $1 \cdot 0$ g cm^{-3}
C $0 \cdot 4$ g cm^{-3}
D $0 \cdot 216$ g cm^{-3}
E $0 \cdot 064$ g cm^{-3}

8.13 In still water a motor boat can travel at 12 miles per hour. The time it will take to travel 12 miles upstream when the current is 4 miles per hour would be

A   40 min
B   45 min
C   60 min
D   80 min
E   90 min

8.14 The current reading on the ammeter A in the following circuit would be

A   1 A
B   3 A
C   4 A
D   9 A
E   36 A

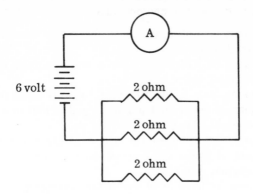

8.15 In a hydraulic press the diameter of one piston is ten times that of the other piston. Compared to the force on the smaller piston, the force on the larger piston will be

A   100 times larger
B   10 times larger
C   the same
D   10 times smaller
E   100 times smaller

8.16 The pressure of water acting on the bottom of a rectangular aquarium, whose base measures 0·15 m by 0·30 m, filled to a depth of 0·10 m is

A   0·45 Nm^{-2}
B   4·50 Nm^{-2}
C   10·0 Nm^{-2}
D   100 Nm^{-2}
E   1000 Nm^{-2}

8.17 An immersion heater supplies heat energy at a constant rate. It will raise the temperature of a certain mass of water from 20°C to 50°C in 10 minutes. It will also raise the temperature of a second liquid from $20^{\circ}C$ to $35^{\circ}C$ in 5 minutes. Neglecting heat losses to the surroundings, you can conclude that the second liquid

A   must be water and have the same mass as the original water
B   must be water and have a smaller mass than the original water
C   must have a specific heat capacity equal to twice that of water and a mass equal to half that of the original water
D   must have a specific heat capacity equal to half that of water and a mass equal to twice that of the original water
E   does not have to be any of the above

8.18   A camera is focussed to take a picture of a girl standing 6 feet from the camera lens. If the film is 4 inches from the lens, then the focal length of the lens expressed in inches is

A $\dfrac{6}{19}$ D $\dfrac{12}{5}$

B $\dfrac{19}{72}$ E $\dfrac{72}{19}$

C $\dfrac{5}{12}$

8.19   Two or more forces can be replaced by a single force that has exactly the same effect. This single force is known as the

A   equilibriant
B   resultant
C   component
D   moment
E   scalar

8.20 The motor of a vacuum cleaner has a rating of 600 watt. The amount of work it can do in 1 minute is

A 1/60 J
B   10 J
C   600 J
D   36 kJ
E 2160 kJ

8.21 In an electroplating experiment, the amount of metal deposited depends on all but two of the following factors:

(i) current
(ii) time
(iii) temperature
(iv) plating metal
(v) separation of the anode and cathode

Which two factors do not affect the amount of metal deposited?

A   (i) and (iv)
B   (ii) and (v)
C   (i) and (iii)
D   (ii) and (iv)
E   (iii) and (v)

8.22 A piano key is struck gently and then struck again, but much harder this time.

A the sound will be louder and the pitch higher than when the key is struck harder
B the sound will be louder when the key is struck harder, but the pitch will not be different
C the sound will be louder when the key is struck harder, but the pitch will be lower
D neither the loudness nor the pitch depend on how hard the key is struck
E none of the above is correct

8.23 A metal washer is heated in a Bunsen flame. It is then dropped into an insulated beaker containing water. If you know the masses of the washer and the water, the initial and final temperatures of the water and the specific heat capacity of water, then without further measurements you could calculate

A the temperature of the flame
B the specific heat capacity of the metal
C the thermal conductivity of the metal
D the specific latent heat of the metal
E none of the above

8.24 A string is vibrating with a frequency of 180 Hz. If the length of the string is made two-thirds as long and nothing else is altered, the frequency of vibration will be

A   80 Hz
B   120 Hz
C   180 Hz
D   270 Hz
E   405 Hz

8.25 Convection can occur in

A gases only
B liquids only
C solids and liquids
D gases and liquids
E solids and gases

8.26 How far away is a thunderstorm in which the thunder is heard 10 seconds after the lightning is seen? The speed of light is 3×10^5 km s^{-1} and the speed of sound is 0.33 km s^{-1}

A   0.066 km
B   2.0 km
C   3.3 km

D 6·6 km

E 6 x 10^4 km

8.27 Sound normally travels fastest in

A a vacuum

B air at low temperature

C air at high temperature

D water

E steel

8.28 Radio waves received by a radio telescope from distant stars may have a wavelength of about 0·20 m. If the speed of the waves is 3 x 10^5 km s^{-1} then the frequency of the waves is

A 6·7 x 10^{-10} Hz

B 6·7 x 10^{-3} Hz

C 1·5 x 10^4 Hz

D 1·5 x 10^6 Hz

E 1·5 x 10^9 Hz

8.29 In a sprint race, the winner must be the runner with the greatest

A maximum acceleration

B average acceleration

C maximum speed

D average speed

E final speed

8.30 Eskimos are able to live comfortably in igloos because, with respect to heat energy, ice is a poor

A radiator

B convector

C conductor

D insulator

E reflector

TEST III

9.1 When a magnet is pushed into a coil of wire, the induced current flows in such a direction as to oppose the change causing it. This illustrates

A Faraday's law
B Lenz's law
C Henry's law
D Gilbert's law
E Fleming's law

9.2 The easiest way to find the wavelength of light ex-experimentally would be to use a

A ripple tank
B diffraction grating
C plane mirror
D glass prism
E stroboscope

9.3 In a theatre, acoustic tiles may be used to

A reflect sound so that everyone can hear
B reflect sound so that there are no 'dead spots'
C absorb sound to prevent wastage of energy
D absorb sound to reduce the reverberation time
E reduce the likelihood of a drop in pitch of speakers' voices

9.4 A voltmeter has a resistance of 1000 ohm and gives full scale deflection when 1 volt is applied across the terminals. To convert the voltmeter so that it shows full scale deflection for 5 volt, which of the following resistors should be used?

A 5000 ohm in series
B 4000 ohm in series
C 200 ohm in series
D 200 ohm in parallel
E 1250 ohm in parallel

9.5 A heavy box is thrust across a rough floor with an initial speed of $4\,\mathrm{m\,s^{-1}}$. It stops moving after 8 seconds. If the resisting force of friction was an average 10 newton, the mass of the box is

A $2 \cdot 5$ kg
B $5 \cdot 0$ kg
C $20 \cdot 0$ kg
D $40 \cdot 0$ kg
E   320   kg

9.6 Which one of these statements concerning the eye is false?

A   the image formed on the retina is the right way up
B   the lens refracts light
C   the focal length of the lens may be varied by the use of the ciliary muscles
D   the distance between the lens and the retina is normally fixed
E   there is a watery fluid in front of the lens

9.7 A parachutist jumps from a plane, opening his parachute a few seconds later. From the moment he jumps to just before he reaches the ground, which one of the following must be constant?

A   speed
B   displacement
C   acceleration
D   velocity
E   none of these

9.8 Complete the following:

$$^{214}_{82}\mathrm{Pb} \longrightarrow {}^{214}_{83}\mathrm{Bi} + \text{gamma ray} + \text{?}$$

A   alpha particle
B   beta particle
C   neutron

D proton

E another gamma ray

9.9 Which one of the following statements about ultrasonic waves is false ?

    A they have greater frequencies than ordinary sound waves

    B they have lower wavelengths than ordinary sound waves

    C they have greater velocities than ordinary sound waves

    D they cannot be heard by humans

    E they may be used in some kinds of cleaning processes

9.10

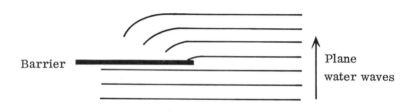

The above diagram shows

    A reflection

    B refraction

    C diffraction

    D interference

    E dispersion

9.11 In the eye of a short-sighted person,

    A the image is formed in front of the retina and a converging lens is necessary to correct this

B   the image is formed behind the retina and a converging lens is necessary to correct this

C   the image is formed in front of the retina and a diverging lens is needed to correct this

D   the image is formed behind the retina and a diverging lens is needed to correct this

E   either a converging or diverging lens is needed to this, depending on the strength of the ciliary muscles

9.12   The velocity ratio of a screw jack is usually

A   much less than 1
B   a little less than 1
C   exactly 1
D   a little more than 1
E   much greater than 1

9.13   To obtain a pure spectrum, it is advisable to pass white light through

A   a prism only
B   a converging lens, a prism, and a second converging lens
C   a diverging lens and a prism
D   two prisms touching each other
E   two prisms close to each other and with two sides parallel

9.14   The boiling point of water may be lowered by

A   adding an impurity
B   reducing the pressure
C   increasing the pressure
D   heating the water more slowly
E   none of the above

9.15 A gas is enclosed in a strong metal container so that the gas cannot escape. The gas is heated from 20°C to 90°C. Its density will

A increase slightly
B increase considerably
C remain the same
D decrease slightly
E decrease considerably

9.16

Eight dry cells, each of e.m.f. 1·5 volts, are supposed to be placed end to end in series. Two of the cells are inadvertently placed upside down as shown. The total e.m.f. from such a set of cells would be

A   0 volt
B   3 volt
C   6 volt
D   9 volt
E   12 volt

9.17 Which of the following is a scalar quantity?

A velocity
B momentum
C force
D mass
E displacement

9.18 The sound heard when you hold a sea-shell close to your ear is due to

A interference
B resonance
C beats

D   total internal reflection

E   diffraction

9.19   When radium decays to radon, an alpha particle is emitted. If radium is represented by $^{226}_{88}$Ra, then radon should be represented as

A $^{222}_{86}$Rn

B $^{224}_{84}$Rn

C $^{226}_{84}$Rn

D $^{222}_{84}$Rn

E $^{226}_{88}$Rn

9.20   On the Kelvin scale, the upper and lower fixed points respectively of a mercury-in-glass thermometer are

A   273 K and 173 K
B   273 K and   0 K
C   373 K and 273 K
D   373 K and   0 K
E   -173 K and -273 K

9.21   Two metal pipes are to be joined tightly together with one of the pipes being inserted a short way into the other as shown.

If the pipes have the same diameter, the best way to join them tightly is to

A   heat the inner pipe to 100°C and then join them
B   heat both pipes to 100°C and then join them
C   cool the outer pipe to -100°C and then join them
D   cool both pipes to -100°C and then join them
E   heat the outer pipe to 100°C and then join them

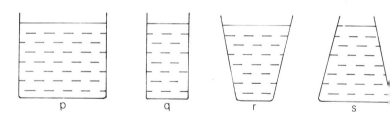

The four vessels above contain different volumes of water. In which vessel will the pressure at the bottom be greatest?

A p

B q

C r

D s

E the pressure will be the same in all the vessels

9.23 A man observes a fish in a pond. The fish appears to be

A deeper in the water than it really is because a ray of light drawn from the fish will refract towards the normal

B deeper in the water than it really is because a ray of light drawn from the fish will refract away from the normal

C nearer the surface than it really is because a ray of light drawn from the fish will refract towards the normal

D nearer the surface than it really is because a ray of light drawn from the fish will refract away from the normal

E at its true depth, unless the man is looking directly down on to the fish

9.24 In a vacuum flask, heat energy transfer due to radiation
is reduced by

A   evacuating the space between the double walls
B   using glass for the double walls
C   supporting the inner flask on cork
D   silvering the surfaces of the double walls
E   using a stopper that fits very tightly

9.25 The mirror used as a shaving mirror is usually

A   convex because this gives a wider field of view
B   concave because this gives a wider field of view
C   convex because this can produce an enlarged image
D   concave because this can produce an enlarged
    image
E   plane because convex or concave mirrors always
    form an inverted image

9.26 How many of the following could occur when water is
heated?

Expansion, contraction, evaporation, boiling, increase
in temperature, increase in mass, increase in density.

A   fewer than 4
B   4
C   5
D   6
E   7

9.27 An object which weighs 10 newton in air is found to
weigh only 4 newton in a liquid. The weight of liquid
it displaces is

A   2·5 N
B   4 N
C   6 N
D   10 N
E   14 N

9.28 A colour television set may have three electron guns which emit electrons that strike the screen. Every point of the screen glows with one of three colours when hit by an electron. In this way a colour picture may be formed. The three colours are

A red, green and yellow
B red, blue and green
C green, orange and blue
D red, blue and yellow
E orange, blue and yellow

9.29 Which of the following is the poorest thermal conductor?

A copper
B paper
C lead
D water
E glass

9.30

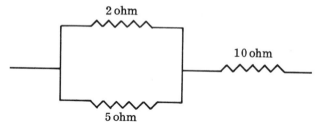

The effective resistance of this combination of resistors is

A $\dfrac{7}{80}$ ohm

B $\dfrac{7}{10}$ ohm

C $\dfrac{70}{17}$ ohm

D $\dfrac{77}{10}$ ohm

E $\dfrac{80}{7}$ ohm

TEST IV

10.1 The temperature of dry ice (solid carbon dioxide) is

    A 373 K
    B between 373 K and 273 K
    C 273 K
    D between 273 K and 0 K
    E 0 K

10.2 The total pressure at the bottom of a liquid does NOT depend on the

    A acceleration due to gravity, g
    B density of the liquid
    C shape of the container
    D depth of the liquid
    E atmospheric pressure

10.3 At the North Magnetic Pole, the angle of dip would be

    A 0°
    B 45°
    C 90°
    D 180°
    E none of these

10.4 A lead-acid accumulator is used in preference to a dry battery in a car because

    A the accumulator can supply a.c.
    B a dry battery cannot supply the necessary high voltage
    C the chemicals used in a dry battery constitute a fire hazard
    D a dry battery would not operate on a cold frosty morning
    E the accumulator can give a bigger current and can be recharged

10.5 The mass of a piece of polonium is found to decrease to a quarter of its original mass of 278 days. The half life of polonium is

A   139 days
B   278 days
C   556 days
D   834 days
E   1112 days

10.6 The working fluid in an aneroid barometer is

A   alcohol
B   mercury
C   water
D   some other liquid
E   none of these

10.7 An electric bell operated within an evacuated bell jar enables us to infer correctly that

A   light waves require a medium for transmission
B   sound waves do not require a medium for transmission
C   light waves do not require a medium for transmission, but sound waves do
D   light waves travel much faster than sound waves
E   light waves and sound waves travel at the same speed

10.8 Two notes are played. If one of them has a frequency of 100 Hz and 10 beats per second are heard, the frequency of the other one is

A   10 Hz
B   10 Hz or 1000 Hz
C   90 Hz or 110 Hz
D   100 Hz
E   1000 Hz

10.9 Using two identical dry cells connected in parallel may be better than using just one because

A   the total power consumption would be reduced

B   the total e.m.f. would be reduced

C   the total e.m.f. would be increased

D   the total current supplied would be greater

E   each cell would be required to produce a smaller current

10.10   Two metal plates placed parallel and close to each other are separated by air thus forming a capacitor. The best way to increase the capacitance would be to

A   move the plates further apart

B   replace the air with argon gas

C   replace the air with a sheet of glass

D   interchange the two plates

E   reduce the area of overlap of the plates

10.11   When an apple falls from a tree, it drops to the ground because of the gravitational attraction between the earth and the apple. If F_1 is the magnitude of the force exerted by the earth on the apple and F_2 is the magnitude of the force exerted by the apple on the earth, then

A F_1 is very much greater than F_2

B F_1 is a little greater than F_2

C F_1 and F_2 are equal

D F_2 is very much greater than F_1

E F_2 is a little greater than F_1

10.12   An object is placed at the principal focus of a converging lens. The image formed is at a distance from the lens of

A   zero

B   less than one focal length

C   one focal length

D   between one and two focal lengths

E   more than two focal lengths

10.13   Suppose you heat some water in a beaker from 20°C to 60°C. Which of the following statements about the water is true ?

A the volume and the density both increase
B the volume and the density both decrease
C the volume increases and the density decreases
D the volume decreases and the density increases
E the volume increases and the density does not change

10.14 Light passing from air to glass is refracted, as is light passing from glass to air. However, when you look out of a window at the view outside, the light does not seem to have been distorted by refraction because

A the angle of refraction is too small to observe
B light incident upon the glass is partially reflected and this tends to mask the effect of refraction
C the emergent ray is parallel to the incident ray and only displacement occurs
D the window pane is too thin for refraction to occur
E the intensity of light outside is far greater than that inside

10.15 Ohm's law tells us that

A the ratio of the current flowing through a metallic conductor to the potential difference across it is called the resistance
B at constant temperature, the ratio of the current flowing through any conductor to the potential difference across it is constant
C the resistance of any conductor at constant temperature is constant
D any conductor has a constant resistance, no matter what the potential difference across it
E at constant temperature, a metallic conductor will have constant resistance

10.16 A photography book has a white cover with a red band around it. When viewed in the darkroom illuminated by red light, the book will appear

A all red
B all black
C white with a red band
D black with a red band
E white with a cyan band

10.17 A simple magnifying glass is formed with

A a diverging lens with the object at a distance from the lens of more than two focal lengths
B a diverging lens with the object at a distance from the lens of between one and two focal lengths
C a converging lens with the object at a distance from the lens of more than two focal lengths
D a converging lens with the object at a distance from the lens of between one and two focal lengths
E a converging lens with the object between the lens and the principal focus

10.18 A climber whose weight is 800 newton climbs a vertical distance of 10 metres in 2 minutes. The amount of work he does is

A 16 kJ
B 8 kJ
C 670 J
D 67 J
E 4 kJ

10.19 A clinical thermometer is usually only marked with temperatures between about 35°C and 43°C. One reason for this is that

A less mercury is needed and the thermometer is cheaper to produce
B only a short temperature range is needed because body temperature does not vary much from 37°C
C expansion of mercury is most uniform within this

range
D the presence of a constriction in the capillary tube does not permit the use of a greater range
E the reading will be unaffected by changes in room temperature

10.20 A hydrometer is an instrument for measuring

    A pressure
    B density
    C temperature
    D humidity
    E altitude

10.21 Compared with the thermal conductivity of ice, the thermal conductivity of snow is

    A smaller because snow contains pockets of air
    B smaller because snow is pure water, but ice will probably contain impurities
    C the same since snow and ice are both forms of water
    D greater because the snow is at a lower temperature
    E greater because snow is moister and this tends to increase conduction

10.22

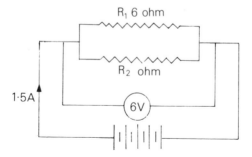

In the above circuit, R_2 is

A 2 ohm

B   4 ohm
C   12 ohm
D   8 ohm
E   6 ohm

10.23   A sound wave in air is an example of

A   an electromagnetic wave
B   a longitudinal wave
C   a transverse wave
D   a shock wave
E   none of the above

10.24   An electric immersion heater that supplies heat energy at a constant rate takes 15 minutes to raise the temperature of some water by 20°C. The same heater takes 30 minutes to raise the temperature of another liquid by 40°C. If the mass of the second liquid is twice that of the water, then the ratio of the specific heat capacity of the second liquid to that of water is

A   1:4
B   1:2
C   1:1
D   2:1
E   4:1

10.25   The spreading out of white light into its component colours is known as

A   refraction
B   deviation
C   dispersion
D   perversion
E   diffraction

10.26   The concept of the bimetallic strip is most likely to be used in designing

A   the balance wheel of a watch
B   an electrical resistance thermometer

C   a pyrometer
D   a lead–acid accumulator
E   a metal tyre to fit a wooden wheel

10.27 At 4°C, a fixed mass of water has its minimum

A   density
B   volume
C   specific heat capacity
D   thermal conductivity
E   temperature

10.28 A polarized wave may be all of the following except

A   a light wave
B   a transverse wave
C   a longitudinal wave
D   a wave that does not vibrate in all directions
E   a wave passing along a rope

10.29

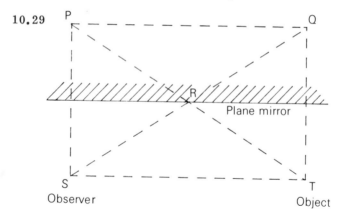

The observer will see a virtual image of the object at

A   point P
B   point Q
C   point R
D   point S
E   point T

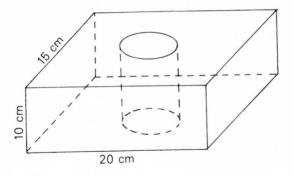

A rectangular solid block has a cylindrical hole bored through it. The mass of the block is now 14·9 kg and the material from which it is made has a density of 5 g cm⁻³. The area of cross-section of the hole is

A   1 cm²
B   2 cm²
C   2·98 cm²
D   10 cm²
E   20 cm²

11.1 Which of the following is a vector quantity?

A mass
B force
C speed
D temperature
E kinetic energy

11.2

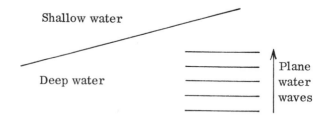

Shallow water

Deep water

Plane water waves

Plane water waves travelling as shown from a region of deep water to shallow water will

A speed up and refract
B speed up, but not refract
C slow down and refract
D slow down, but not refract
E refract, but continue at the same speed

11.3 An 800 kg car travelling at $25\,\text{m s}^{-1}$ is involved in a collision and is brought to a halt in just 10 seconds. The average retarding force must be about

A 320 N
B 1000 N
C 2000 N
D 10000 N
E 20000 N

11.4 A child weighing 25 kg slides down a rope hanging from a branch of a tall tree. If the force of friction acting against him is 200 N, what is the child's acceleration? If necessary, assume that the acceleration due to gravity, g, is about $10 \, \text{m s}^{-2}$.

A $2 \, \text{m s}^{-2}$
B $5 \, \text{m s}^{-2}$
C $8 \, \text{m s}^{-2}$
D $22 \cdot 5 \, \text{m s}^{-2}$
E $50 \, \text{m s}^{-2}$

11.5   In a camera, the function of the film is most like the function of which part of the eye?

A   the pupil
B   the schlerotic layer
C   the retina
D   the cornea
E   the iris

11.6   If alpha, beta and gamma rays are arranged in order of penetrating power, starting with the most penetrating, the most usual order would be

A   alpha, gamma, beta
B   alpha, beta, gamma
C   beta, alpha, gamma
D   gamma, alpha, beta
E   gamma, beta, alpha

11.7   A motor is operated by four $1 \cdot 5$ V cells connected in parallel. If the resistance of the motor is 10 ohm, then the current drawn is

A $0 \cdot 0375$ A
B $0 \cdot 15$ A
C $0 \cdot 60$ A
D $1 \cdot 67$ A
E $6 \cdot 67$ A

11.8 A string is vibrating with a frequency of 180 Hz. To double this frequency by adjustment only of the tension in the string, would require the tension to be made

A four times as great
B twice as great
C three times as great
D one half as great
E one fourth as great

11.9 When an electric current is passed through distilled water which has been acidified with a little sulphuric acid, the gasses collecting at the electrodes are

	Cathode		Anode
A	2 vol. hydrogen	:	1 vol. oxygen
B	1 vol. hydrogen	:	1 vol. oxygen
C	1 vol. hydrogen	:	2 vol. oxygen
D	2 vol. oxygen	:	1 vol. hydrogen
E	1 vol. oxygen	:	2 vol. hydrogen

11.10 An object painted red is viewed under green light. The object appears

A red
B green
C yellow
D black
E white

11.11 The half life of a radioactive material is

A how much material is left when half has decayed
B how much material has decayed when half is left
C how long it takes for half the material to decay
D how much more time will be needed for complete decay after half has already decayed
E half the time it takes for all the material to decay

11.12 An object is thrown up into the air with an initial speed of 5 m s^{-1}. If the acceleration due to gravity, g, is 10 m s^{-2}, the object will return to the ground after a total time of

A 0·5 s
B 1·0 s
C 1·5 s
D 2·0 s
E 3·0 s

11.13 The bob of a pendulum is in the form of a flat copper disc. If this is allowed to swing so that it passes between the poles of a powerful magnet, it will quickly come to rest because

A the copper becomes magnetised, since the magnet is a very powerful one, and the copper is attracted
B a current is induced in the disc and the magnetic field associated with this current tends to oppose motion
C the force exerted by the magnet is far greater than the force of gravity and therefore the disc cannot swing
D the pendulum becomes electrically charged as it swings through the air - especially on a hot day - and thus is strongly attracted to the magnet
E just as the copper is an excellent thermal and electrical conductor, it can conduct the lines of force and the disc stops swinging in accordance with Lenz's law

11.14 A glass flask is partly filled with water which is then heated over a Bunsen flame. As soon as the water begins to boil the flask is removed from the Bunsen flame and tightly stoppered immediately. Boiling stops. When some cold water is poured onto the outside of the flask, the water inside will

A begin to boil since contraction of the flask reduces the pressure inside

B begin to boil since condensation of steam reduces the pressure inside

C not boil since its temperature is less than 100°C

D not boil since contraction of the flask increases the pressure inside

E not boil since condensation of steam increases the pressure inside

11.15 The main reason why a pinhole camera works is the fact that light

A is refracted as it passes through a lens

B travels in straight lines

C is reflected when it strikes any object

D diffracts around objects

E travels at very high speeds

11.16 Which one of these statements about the resistance of a wire is false?

A it may be measured in units of ohm

B it depends on what the wire is made from

C it is inversely proportional to the diameter of the wire

D it is directly proportional to the length

E it depends on the temperature

11.17 The focal length of a concave mirror is 12 cm. To form a virtual image that is twice as far from the mirror as the object is, the object should be placed at a distance from the lens of

A 4 cm

B 6 cm

C 8 cm

D 10 cm

E 12 cm

11.18 If gamma rays, visible light and radio waves are arranged in order of increasing wavelength, starting with the shortest wavelength, the correct order would be

A gamma, light, radio
B gamma, radio, light
C light, radio, gamma
D radio, gamma, light
E radio, light, gamma

11.19 Resistance is best measured with a

A resistance thermometer
B Wheatstone bridge
C galvanometer
D voltameter
E sonometer

11.20 When an object is placed 10 cm from a lens, the virtual image formed is 5 cm from the lens. This means that the type of lens and the focal length are

A converging: 3·3 cm
B converging: 10 cm
C converging: 15 cm
D diverging: 3·3 cm
E diverging: 10 cm

11.21 It is possible to boil water in a paper cup held over a Bunsen flame. To ensure a successful demonstration, it is best to use

A thin paper and a large base for the cup
B thin paper and a small base for the cup
C thick paper and a large base for the cup
D thick paper and a small base for the cup
E a very small volume of water

ANSWERS

1.1 E	2.1 A	3.1 E	4.1 D
1.2 C	2.2 D	3.2 B	4.2 E
1.3 C	2.3 C	3.3 C	4.3 E
1.4 A	2.4 B	3.4 D	4.4 C
1.5 B	2.5 C	3.5 D	4.5 B
1.6 C	2.6 B	3.6 D	4.6 C
1.7 D	2.7 D	3.7 B	4.7 C
1.8 D	2.8 C	3.8 C	4.8 A
1.9 E	2.9 A	3.9 B	4.9 C
1.10 B	2.10 C	3.10 D	4.10 B
1.11 A	2.11 D	3.11 E	4.11 C
1.12 C	2.12 A	3.12 A	4.12 E
1.13 D	2.13 B	3.13 D	4.13 C
1.14 C	2.14 C	3.14 C	4.14 C
1.15 C	2.15 C	3.15 D	4.15 B
1.16 B	2.16 B	3.16 B	4.16 A
1.17 B	2.17 C	3.17 C	4.17 B
1.18 B	2.18 D	3.18 D	4.18 D
1.19 C	2.19 C	3.19 A	4.19 B
1.20 C	2.20 B	3.20 B	4.20 B
1.21 D	2.21 B	3.21 C	4.21 E
1.22 B	2.22 C	3.22 C	4.22 D
1.23 A	2.23 A	3.23 C	4.23 A
1.24 D	2.24 C	3.24 A	4.24 D
1.25 B	2.25 B	3.25 C	4.25 C
1.26 B	2.26 E	3.26 E	4.26 C
1.27 A	2.27 C	3.27 E	4.27 A
1.28 D	2.28 D	3.28 E	4.28 C
1.29 B	2.29 E	3.29 A	4.29 A
1.30 E	2.30 A	3.30 D	4.30 A

5.1 B	6.1 B	7.1 D	8.1 D
5.2 C	6.2 B	7.2 E	8.2 B
5.3 D	6.3 D	7.3 D	8.3 B
5.4 D	6.4 C	7.4 B	8.4 E
5.5 A	6.5 E	7.5 A	8.5 E
5.6 C	6.6 A	7.6 A	8.6 B
5.7 E	6.7 D	7.7 E	8.7 A
5.8 C	6.8 C	7.8 C	8.8 B
5.9 A	6.9 C	7.9 A	8.9 C
5.10 D	6.10 B	7.10 B	8.10 C
5.11 C	6.11 C	7.11 D	8.11 D
5.12 D	6.12 A	7.12 B	8.12 A
5.13 A	6.13 B	7.13 E	8.13 E
5.14 B	6.14 C	7.14 C	8.14 D
5.15 D	6.15 D	7.15 C	8.15 A
5.16 C	6.16 C	7.16 B	8.16 E
5.17 A	6.17 B	7.17 C	8.17 E
5.18 C	6.18 C	7.18 B	8.18 E
5.19 C	6.19 C	7.19 D	8.19 B
5.20 B	6.20 C	7.20 D	8.20 D
5.21 B	6.21 A	7.21 B	8.21 E
5.22 D	6.22 D	7.22 B	8.22 B
5.23 C	6.23 D	7.23 C	8.23 E
5.24 C	6.24 B	7.24 B	8.24 D
5.25 C	6.25 D	7.25 B	8.25 D
5.26 C	6.26 C	7.26 D	8.26 C
5.27 A	6.27 C	7.27 B	8.27 E
5.28 D	6.28 D	7.28 A	8.28 E
5.29 D	6.29 E	7.29 B	8.29 D
5.30 D	6.30 E	7.30 C	8.30 C

9.1 B	10.1 D	11.1 B
9.2 B	10.2 C	11.2 C
9.3 D	10.3 C	11.3 C
9.4 B	10.4 E	11.4 A
9.5 C	10.5 A	11.5 C
9.6 A	10.6 E	11.6 E
9.7 E	10.7 C	11.7 B
9.8 B	10.8 C	11.8 A
9.9 C	10.9 E	11.9 A
9.10 C	10.10 C	11.10 D
9.11 C	10.11 C	11.11 C
9.12 E	10.12 E	11.12 B
9.13 B	10.13 C	11.13 B
9.14 B	10.14 C	11.14 B
9.15 D	10.15 B	11.15 B
9.16 C	10.16 A	11.16 C
9.17 D	10.17 E	11.17 B
9.18 B	10.18 B	11.18 A
9.19 A	10.19 B	11.19 B
9.20 C	10.20 B	11.20 E
9.21 E	10.21 A	11.21 A
9.22 E	10.22 C	11.22 C
9.23 D	10.23 B	11.23 B
9.24 D	10.24 B	
9.25 D	10.25 C	
9.26 D	10.26 A	
9.27 C	10.27 B	
9.28 B	10.28 C	
9.29 B	10.29 B	
9.30 E	10.30 B	